JUMBO COLORING BOOK

COMMON NAME: BEES AND HORNETS

Hymenoptera is one of the largest and most well-known orders of insects, and it includes Bees and Hornets.

Biographical Note
The Nature of Bees & Hornets An Educational Coloring Book is a new work,
first published by Little Artist Studio in 2025.

International Standard Book Number
ISBN 979-8-9992504-5-2

www.littleartiststudio.org

Celebrate the power and purpose of pollinators with over 90 stunning coloring pages featuring bees and hornets from around the world. These remarkable insects are vital to healthy ecosystems, playing key roles in pollination, pest control, and biodiversity. Part of Little Artist Studio's acclaimed educational coloring series, each full-page illustration is designed to spark curiosity, inspire learning, and encourage artistic expression. With single-sided pages, artists of all ages can use any medium and proudly display their finished creations. Perfect for nature lovers, educators, and budding entomologists alike.

SCIENTIFIC NAME FOR BEES: APOIDEA (SUPERFAMILY)

Apoidea is the scientific name for large group that includes all bees, helping scientists identify which insects belong to this group.

SCIENTIFIC NAME FOR HONEY BEES: APIS MELLIFERA

- Apis mellifera is the scientific name for honey bees, which means "bee that makes honey."

SCIENTIFIC NAME FOR HORNETS: VESPA (GENUS OF TRUE HORNETS)

Vespa is the special science name for hornets, helping scientists identify which insects are hornets.

LEARNING ABOUT BEES AND HORNETS

Entomology is the study of insects. Bees and hornets are insects too. Bees make honey and help flowers grow, while hornets build nests and protect themselves by stinging.

Bees

Hornets

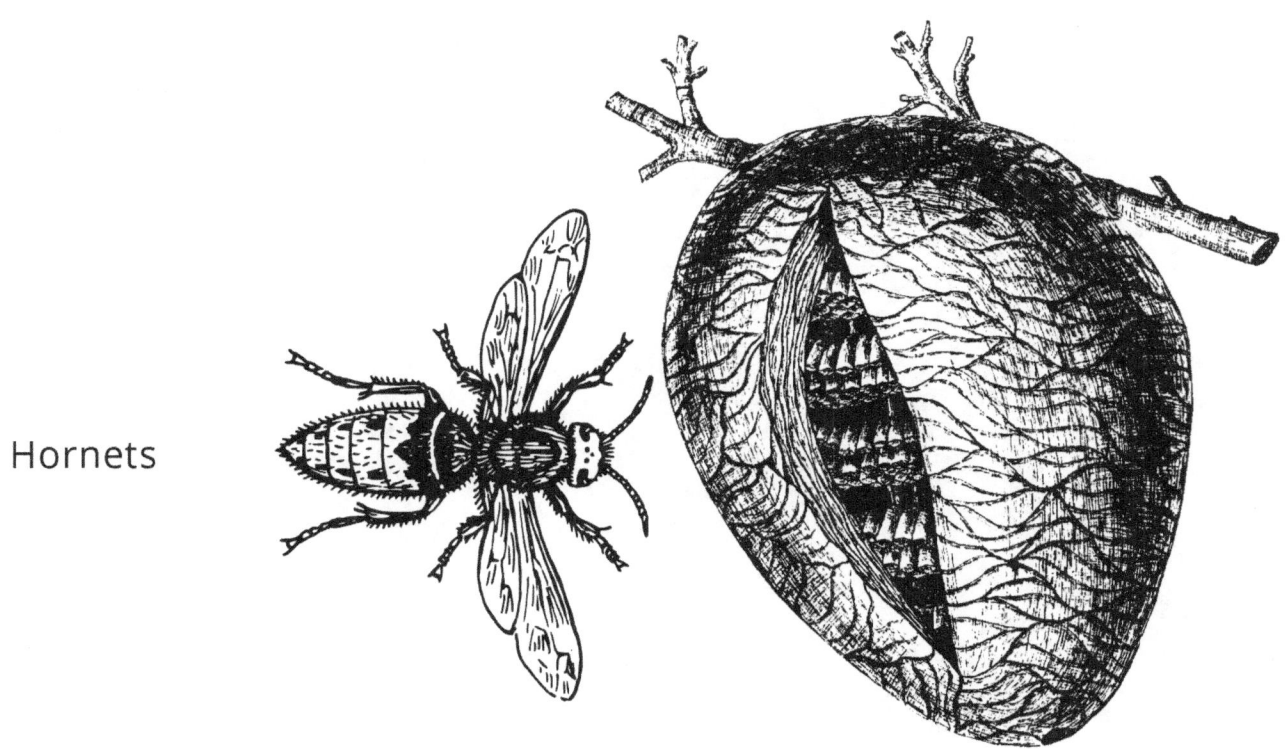

BEES AND HORNETS: EGG STAGE

Even though they look and act differently, the life cycles of bees and hornets follow the same basic steps!

The queen bee's job is to lay all the eggs in the hive.

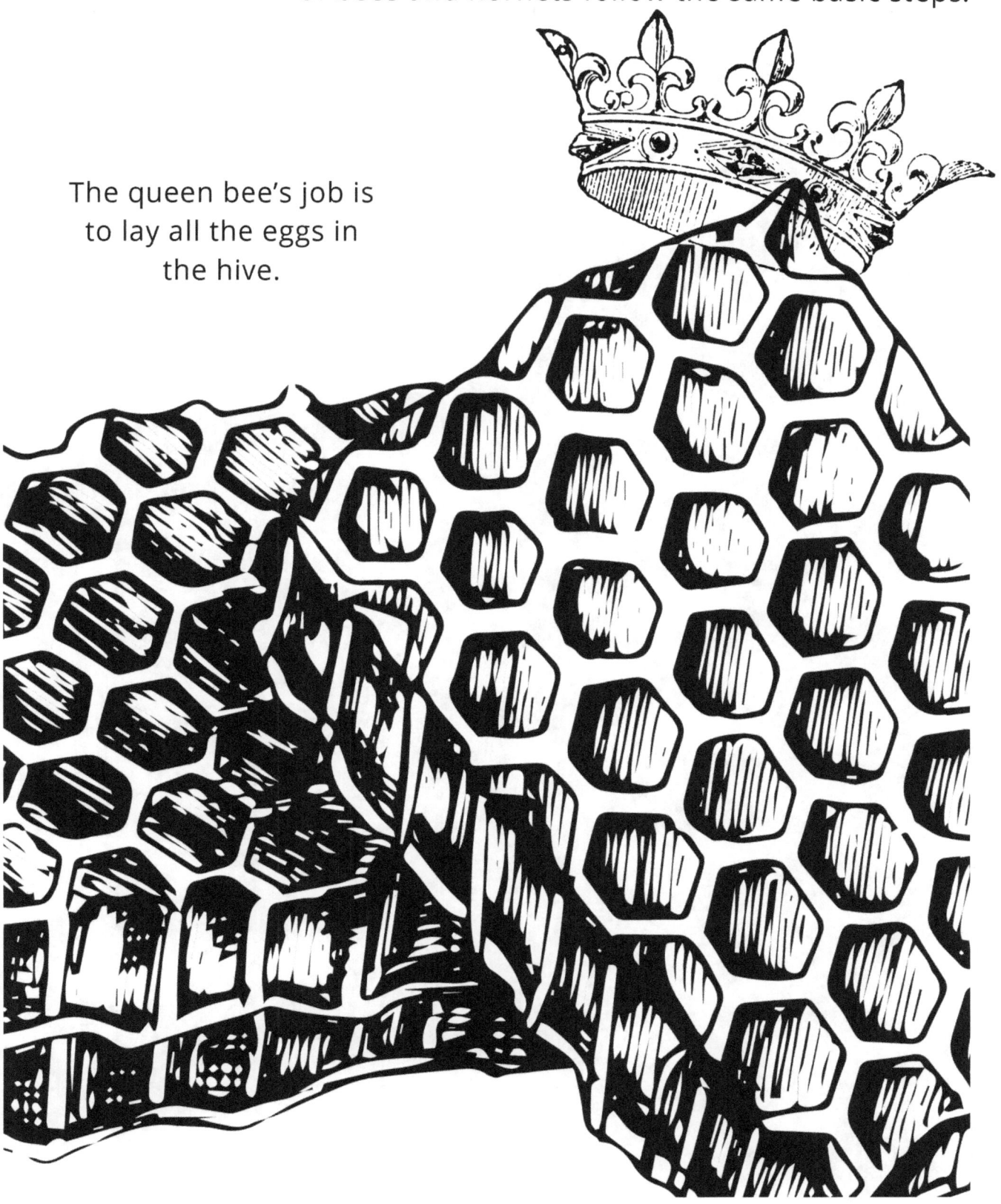

BEES AND HORNETS: LARVA STAGE

The eggs hatch into larvae, which grow by eating food.

BEES AND HORNETS: PUPA STAGE

The larvae become pupae and begin changing into adult insects.

BEES AND HORNETS: ADULT STAGE

The adult bee or hornet comes out, ready to work or fly.

Bees

Hornets

SPECIAL FEATURE: BEES

Bees have special hairy bodies that help them collect pollen from flowers.

SPECIAL FEATURE: HORNETS

Hornets can sting many times without dying, and they use their powerful sting to protect their nest.

Stinger

BEES

Bees are amazing pollinators and help plants grow fruits and vegetables.

BEES

Some bees even have special "pollen baskets" on their legs to carry pollen home!

HORNETS

Hornets chew up wood to build their nests, making a paper-like material!

BEES

Honey bees make honey by turning flower nectar into a sweet, sticky food.

BEES

Bees can communicate with each other by doing a special "waggle dance" to show where flowers are.

HORNETS

Hornets can sting multiple times to protect their nest.

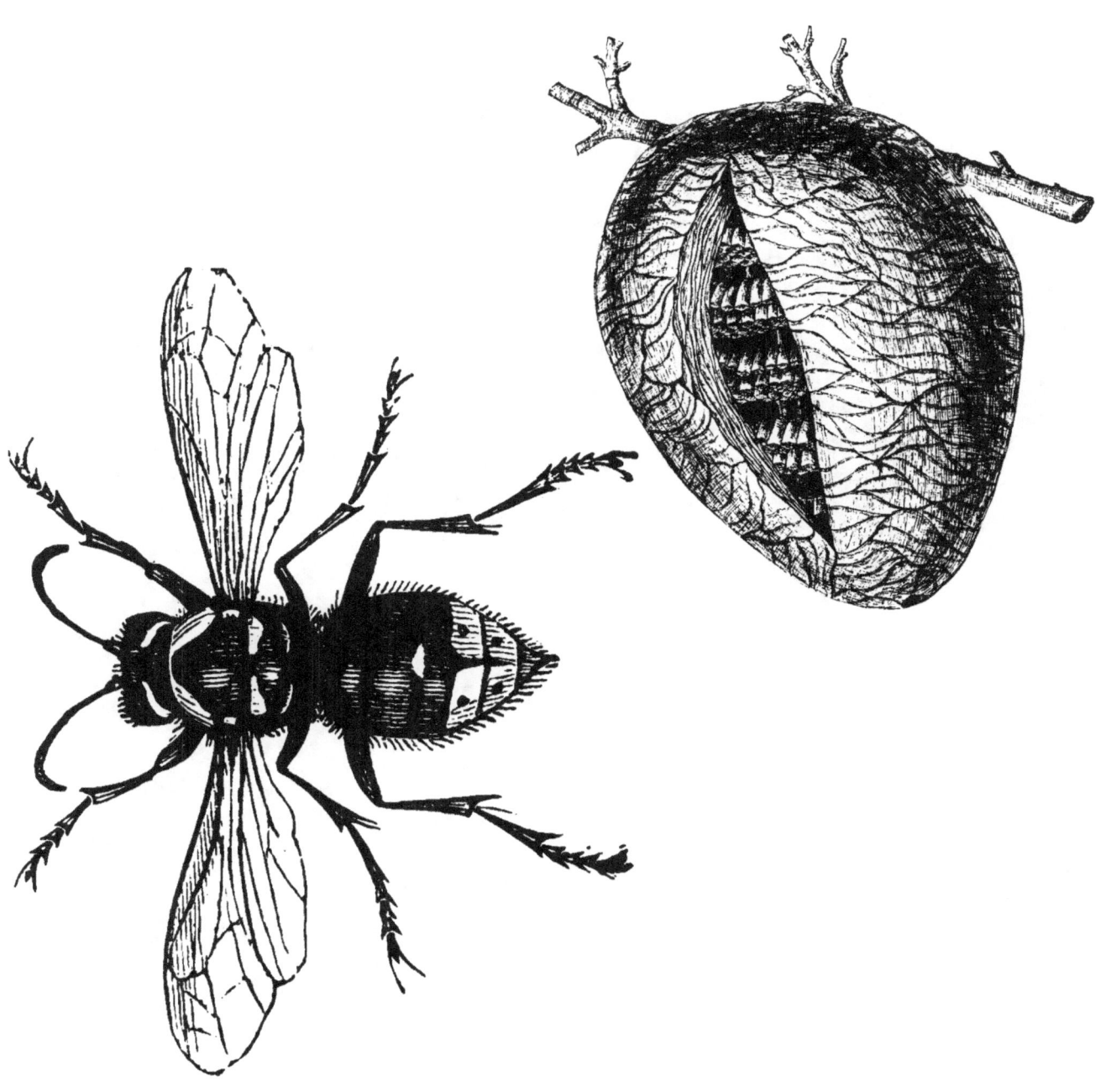

HORNETS

Hornets help control pest bugs by hunting and eating them.

BEES

Bees build their nests out of wax that they make from their own bodies!

BEES

Inside a hive, bees live in a big family with a queen, workers, and drones - all working together!

BEES

Honey bees have special glands that produce wax, which they shape into hexagon-shaped honeycombs - perfect for storing honey, pollen, and eggs.

HORNETS

Hornets are not a separate species from wasps-they're actually a subgroup of wasps in the genus Vespa. They tend to be larger and more aggressive than common wasps.

HONEY BEES

Honey bees collect nectar from flowers and turn it into honey to eat and store for winter.

BEES NEST

Bumblebees may build a nest underground or in hidden places like old logs or walls.

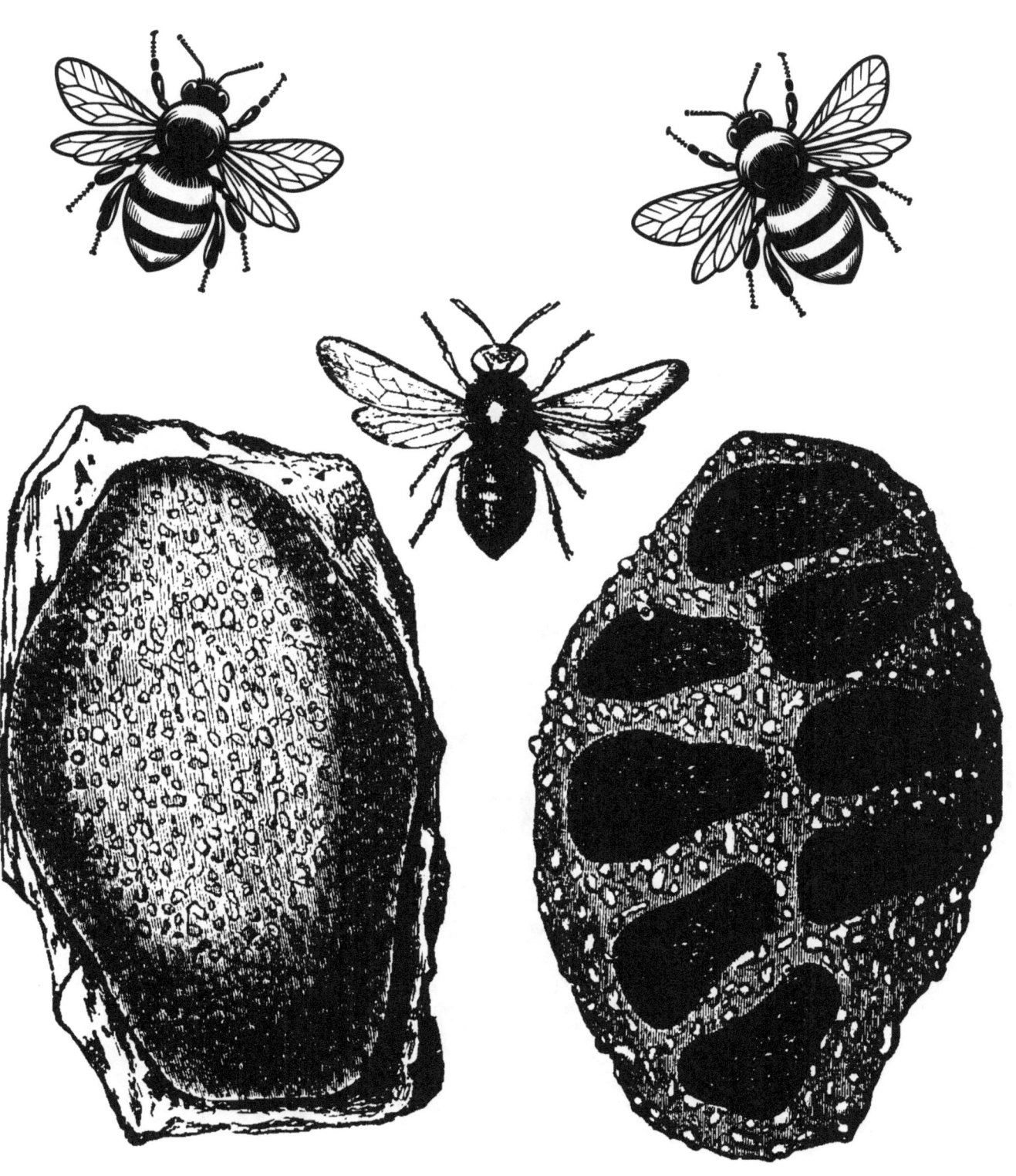

BEEHIVE

Honey bees live in a hive, which can be made in tree hollows or man-made boxes (like beekeepers use).

BEE EYES

Bees have hairy eyes! Those tiny hairs help them collect pollen — it sticks like velcro!

QUEEN BEE

The queen bee is the hive's sole reproductive female, responsible for laying up to 2,000 eggs a day to produce worker bees and drones - making her the ultimate multitasker of the colony.

DRONE BEES

A drone bee is the larger, stingless male honeybee whose only role in the hive is to mate with the queen.

WORKER BEE

Worker bees are female honeybees that lack reproductive capacity and perform essential tasks within the hive.

BEES

BEEHIVE

A single beehive can hold up to 60,000 bees - and every bee has a special job!

BEES AND FLOWERS

BEEKEEPER

LANGSROTH HIVE

Beekeepers use a special little house called a Langstroth hive. It's where bees live, make honey, and take care of baby bees.

SMOKER

Beekeepers use a tool called a smoker. It blows out cool smoke that helps keep the bees calm while the beekeeper checks the hive

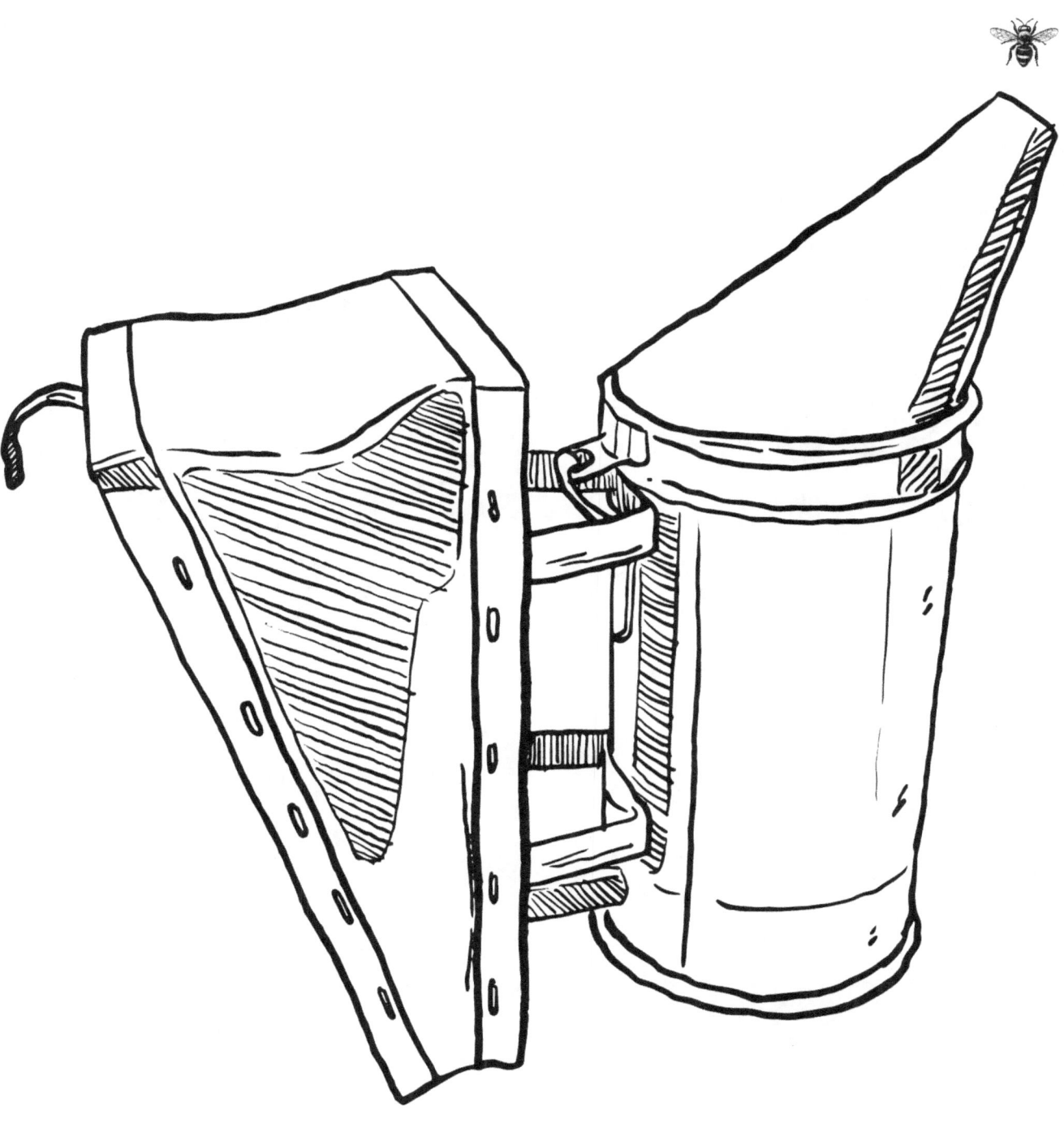

SWARM OF BEES

BEES MAKE HONEY

Honey has natural healing powers! It can kill germs, reduce swelling, and help cuts and burns heal faster. People have used honey as medicine for thousands of years.

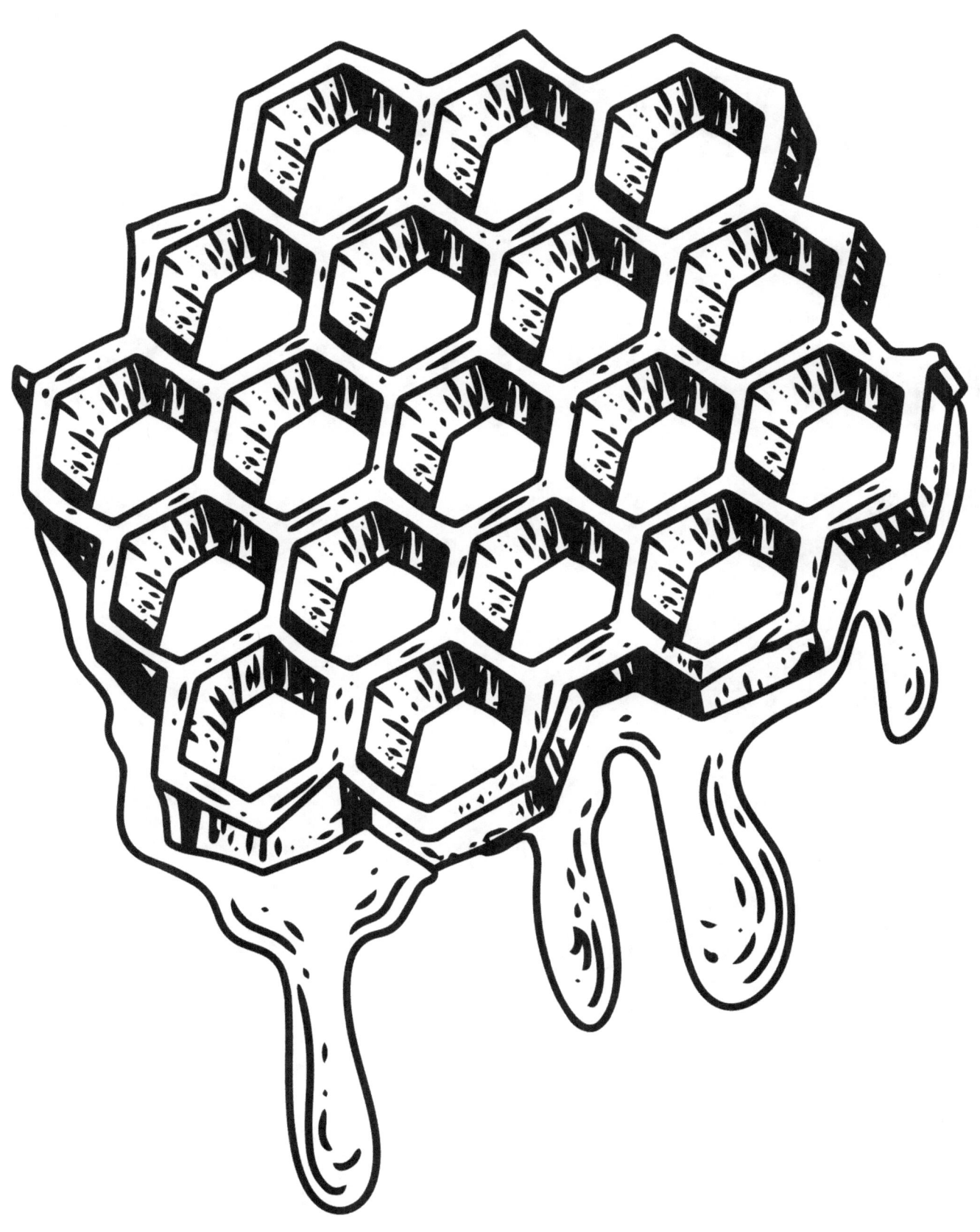

BEE ECONOMICS

Bee economics is the study of how bees help the economy, especially through their work as pollinators.

BEE ECONOMICS

Pollination is when bees move pollen from one flower to another, helping plants grow fruits, vegetables, and seeds.

BEES BUZZ

The buzzing sound comes from their wings flapping 200 times every second. It's like tiny engines helping them zoom through the air!

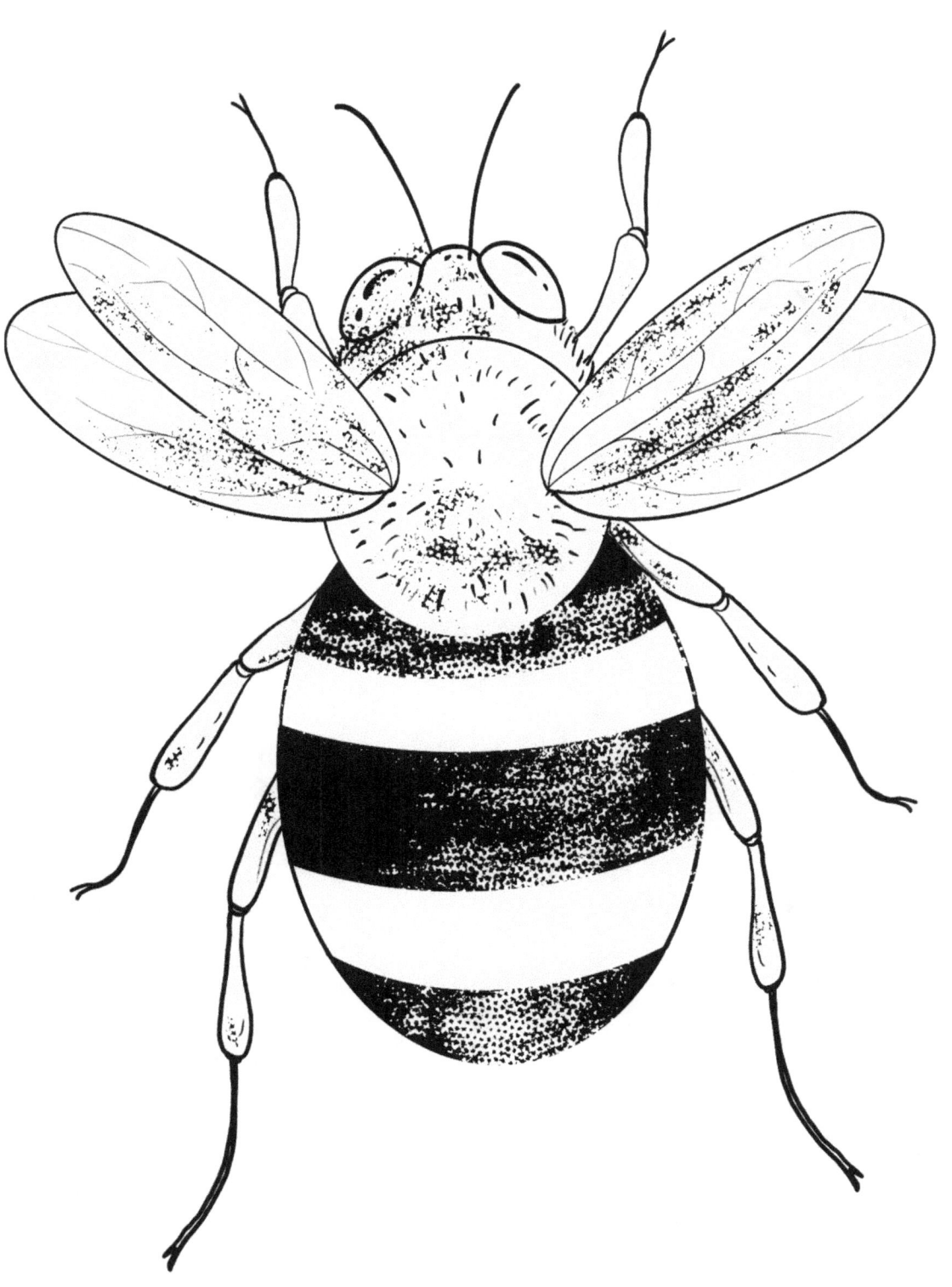

FLOWER POWER

Bees help plants grow by moving pollen from flower to flower.

BEES

While bees aren't the fastest insects, they're super strong flyers and can visit hundreds of flowers in one trip!

COLONIES

Honey bees live in large groups called colonies.

BEEHIVE

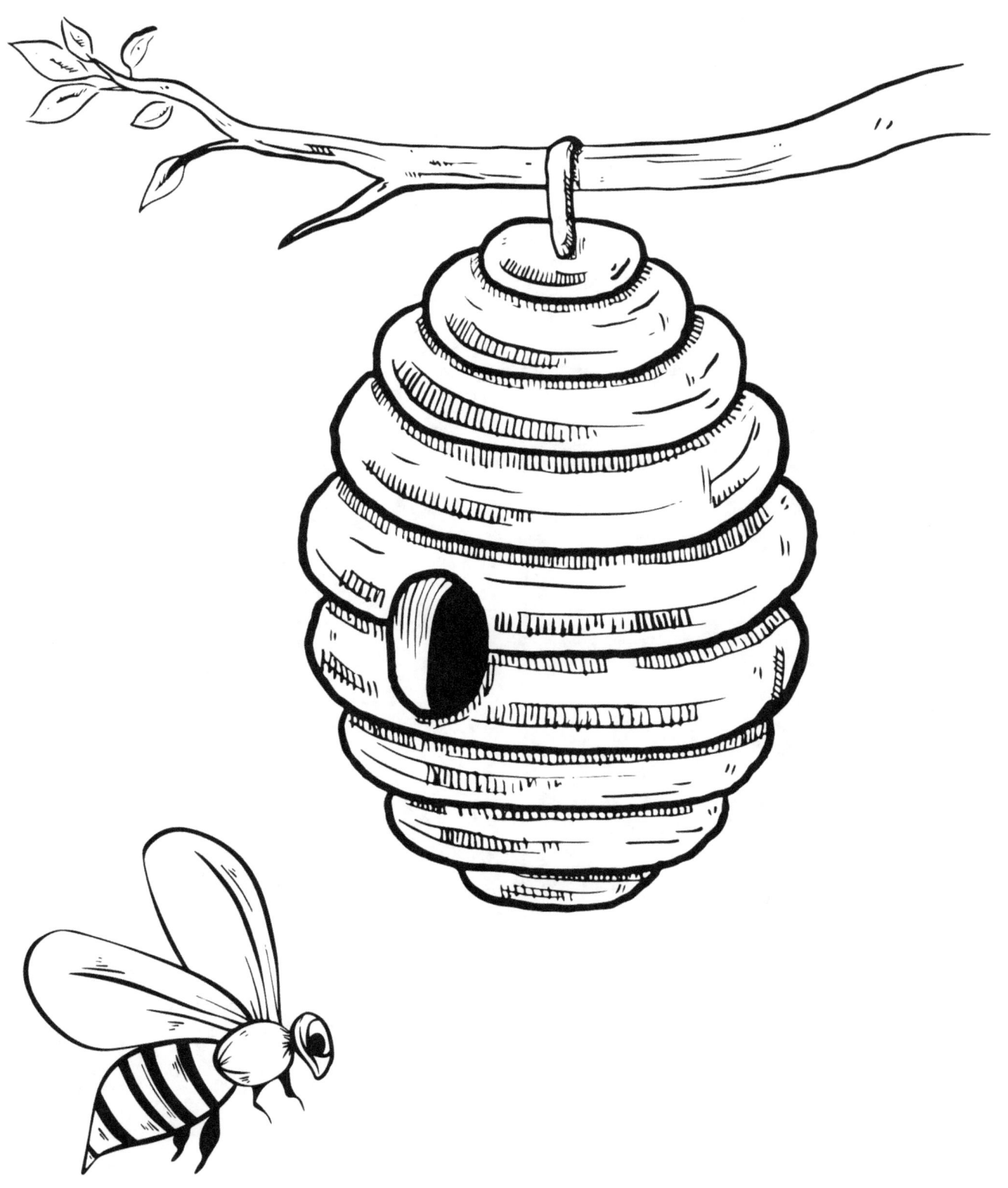

HEXAGONS

Bees build their hive out of perfect hexagons—nature's most efficient shape!

WAGGLE DANCE

Honey bees do a "waggle dance" to tell other bees where to find flowers.

BUZZING AROUND

Bees can fly up to 15 miles per hour (about 24 kilometers per hour)!

HONEY BEES

One honey bee makes only about 1/12 of a teaspoon of honey in its whole life!

HONEY BEES

It takes about 12 bees to make just one teaspoon of honey!

BEES AND BEARS

Yes, bears eat honey and are attracted to beehives!

HORNETS

Hornets will fiercely defend their nests if they sense a threat. Vibrations, movement, or even getting too close can provoke an attack.

HORNETS

Hornets chew wood fibers and mix them with saliva to create a paper-like material. This is what they use to build their large, layered nests.

ASIAN GIANT HORNET

The Asian giant hornet (Vespa mandarinia), also known as the "murder hornet," is the world's largest hornet, reaching up to 2 inches (5 cm) in length.

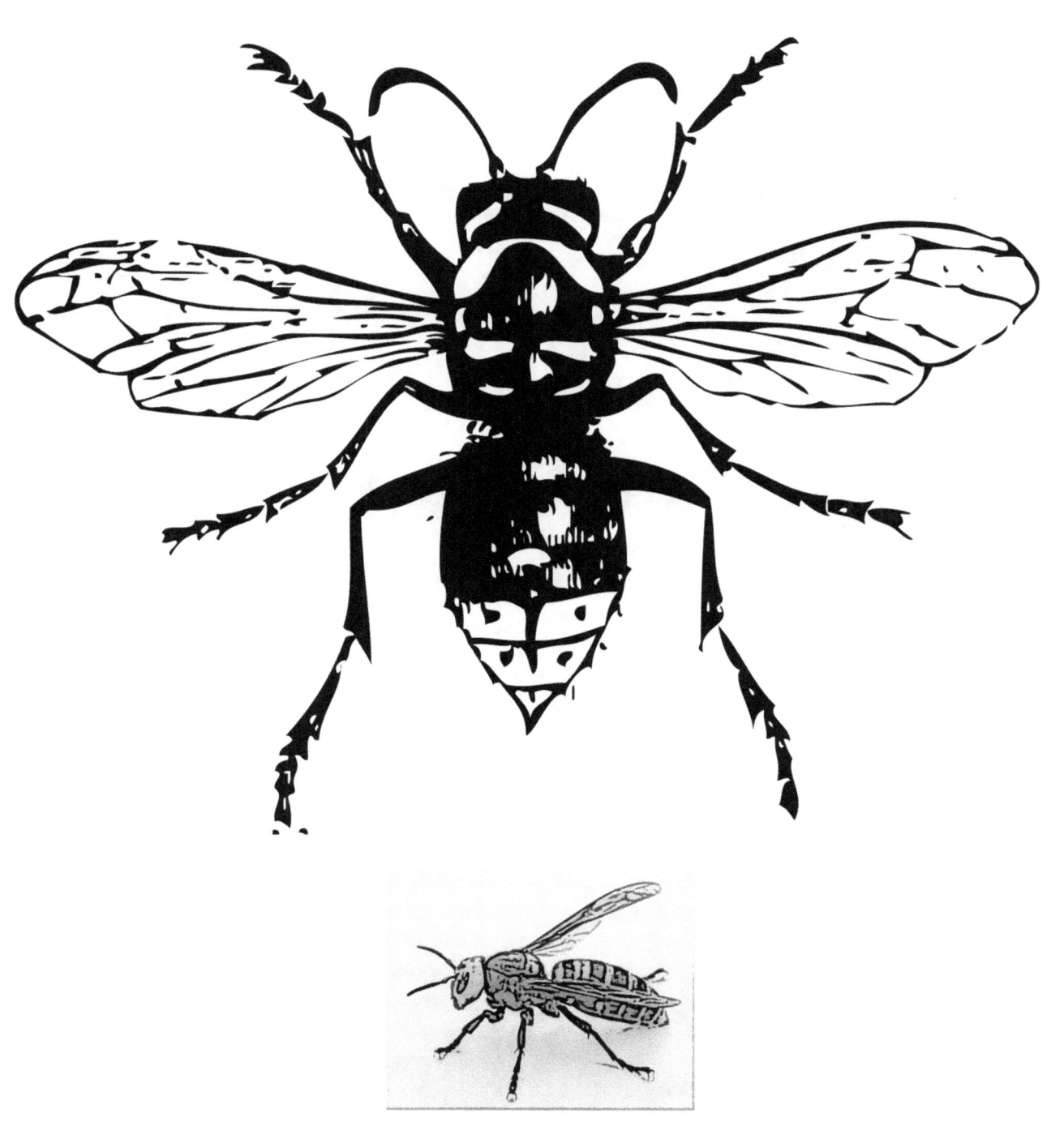

ASIAN GIANT HORNET

The Asian giant hornet can wipe out a whole beehive in just a few hours, using its sharp jaws - that's why bees are so scared of it! .

HORNET VENOM

Hornet stings inject venom that can cause intense pain, swelling, and even allergic reactions.

HORNET STING

Unlike bees, hornets can sting multiple times.

HORNET SPECIES

Native hornet species are found across Europe and Asia.

HORNET SPECIES

Some hornets like the Asian hornet (Vespa velutina), have made their way to other continents invading parts of Europe and the U.S.

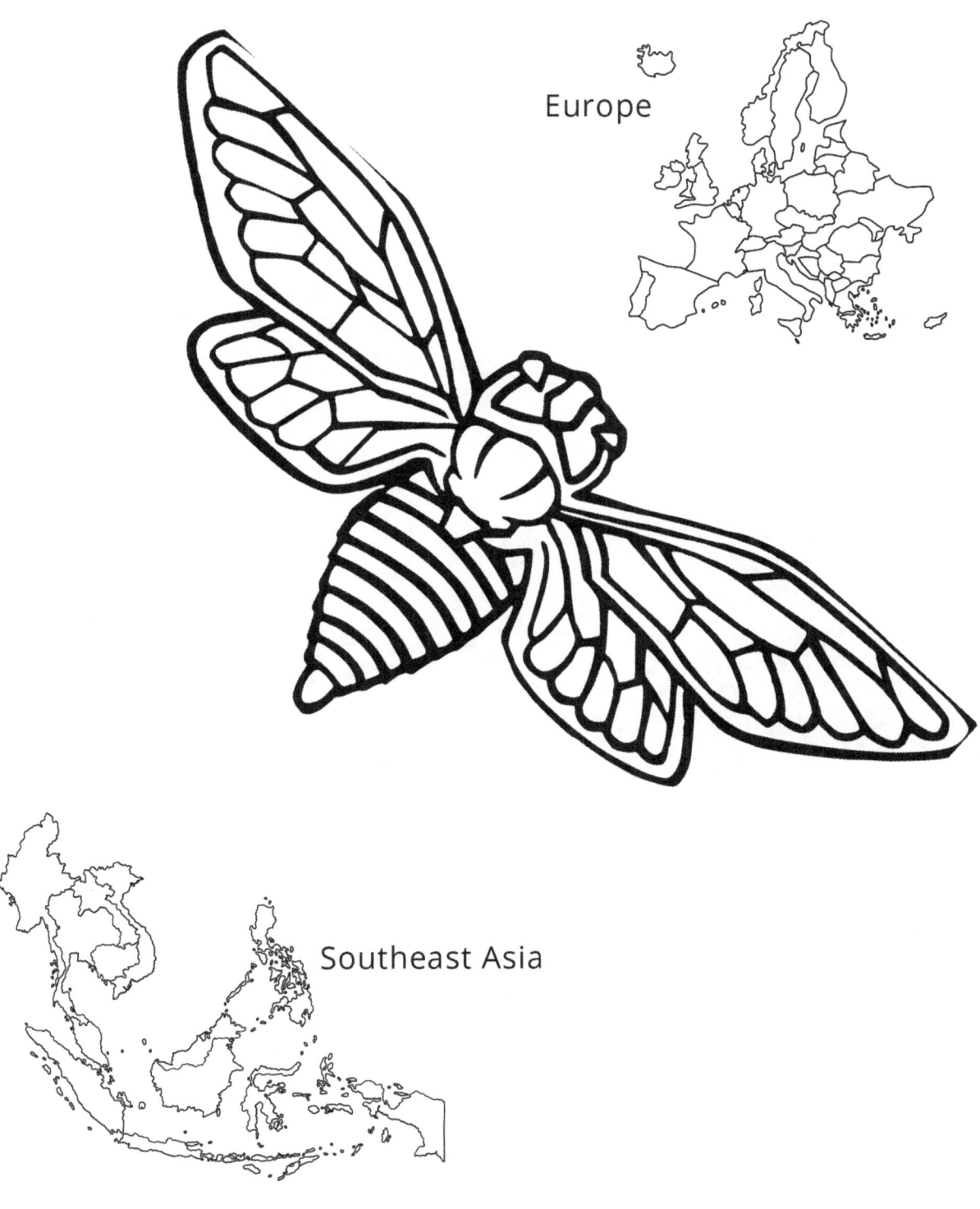

Europe

Southeast Asia

HORNETS NEST

If a colony survives the winter (which is rare), hornets can sometimes reuse and add to their old nest, making it even larger the following season.

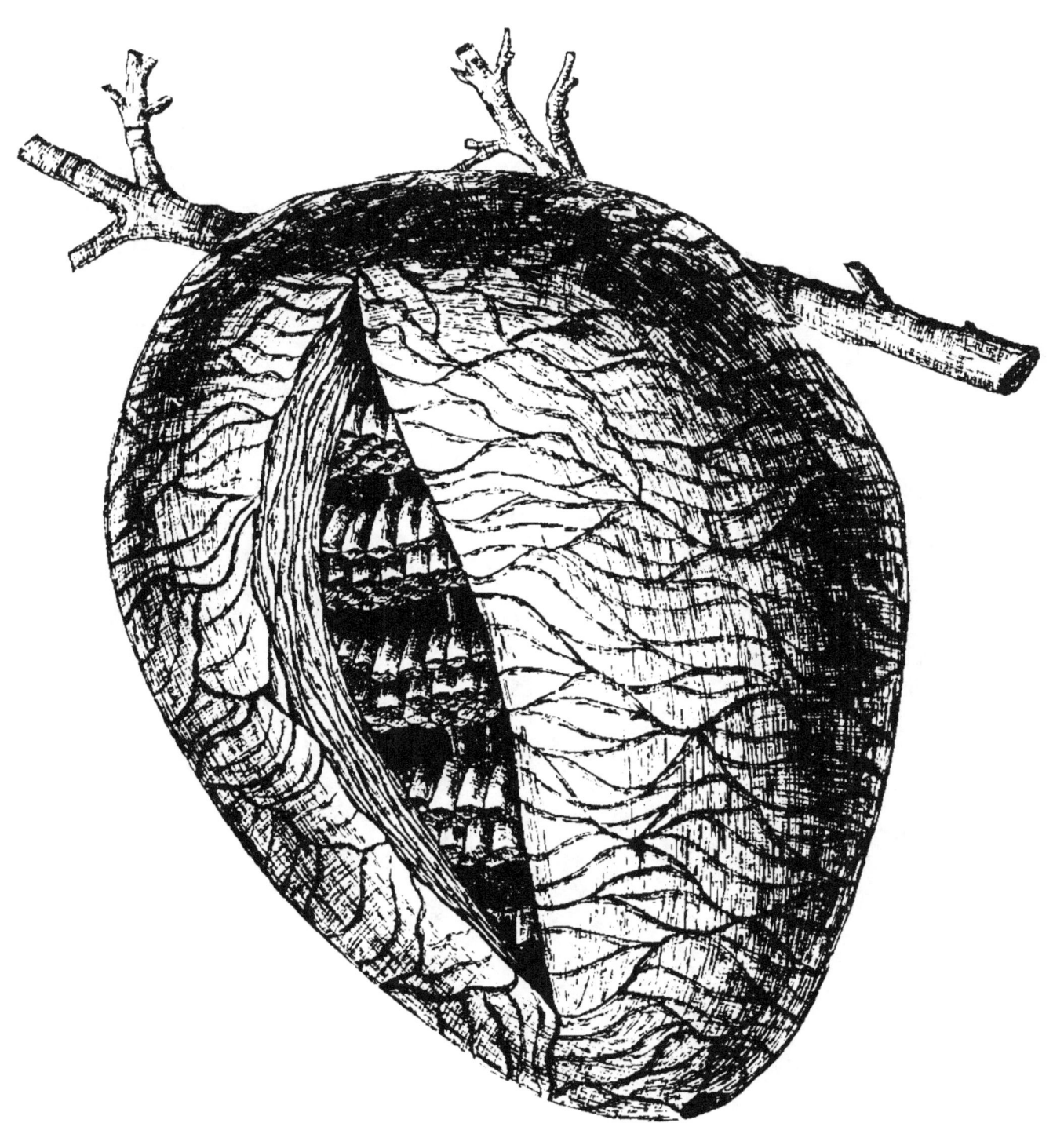

HORNET STING

When one hornet stings, it releases a pheromone that alerts other hornets to join in the attack. This is why hornet attacks can escalate quickly.

HORNETS

Hornets are carnivorous predators

Caterpillar

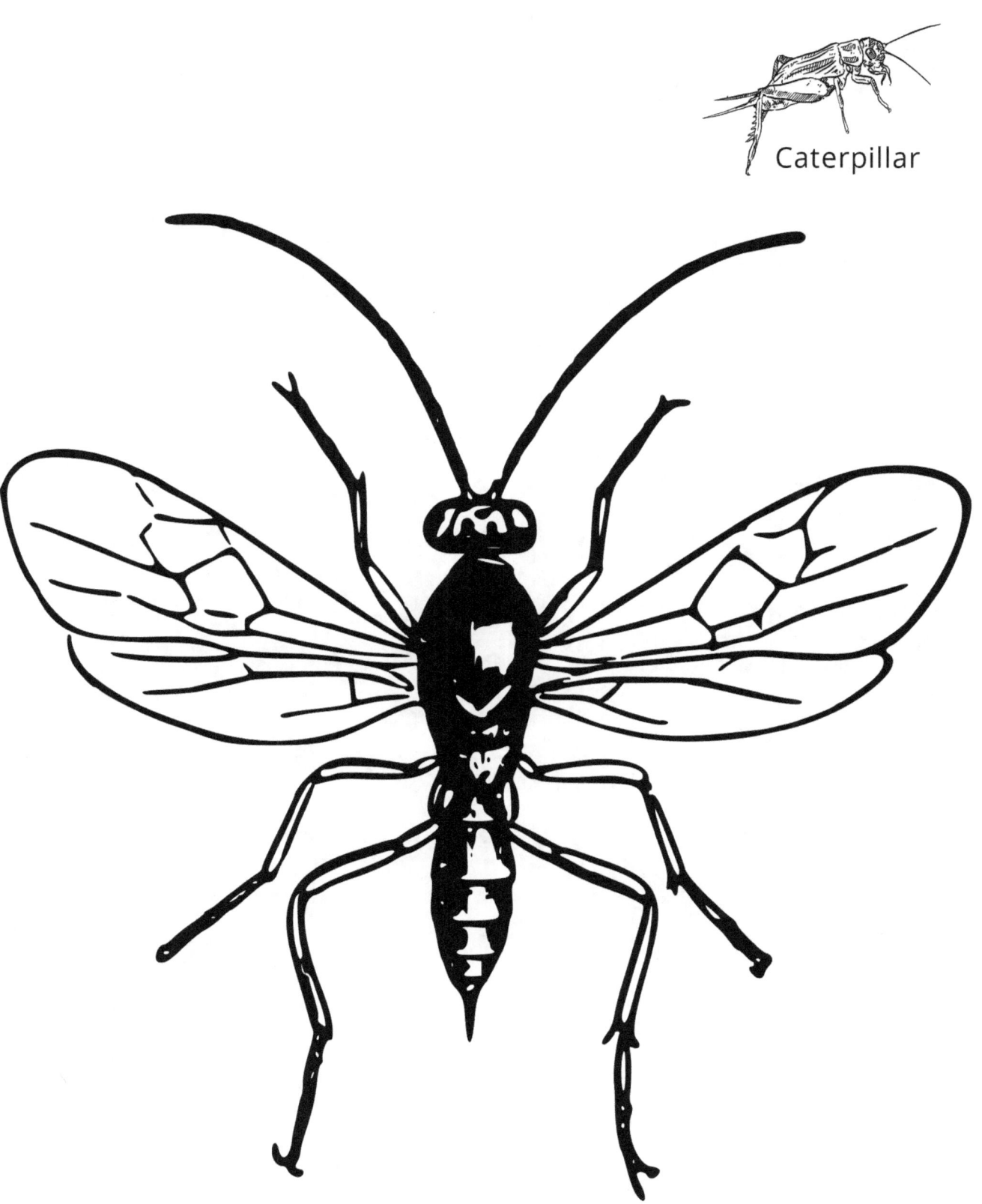

HORNET HUNTERS

Hornets hunt other insects like flies, caterpillars, and even bees. They chew them up and feed the protein-rich paste to their larvae.

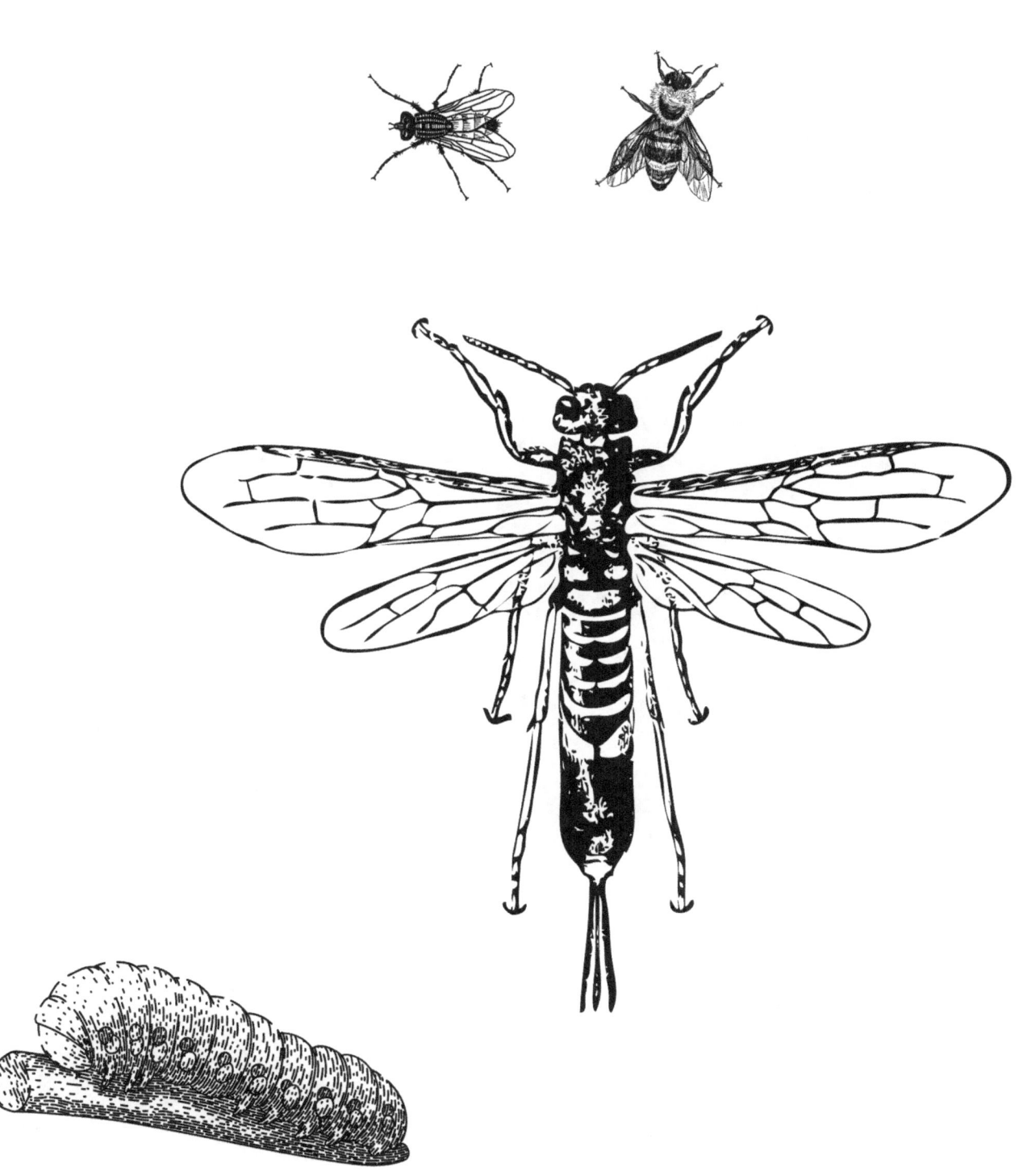

HORNET HUNTERS

Hornets are also predators, hunting and feeding their larvae on insects like grasshoppers, crickets, and caterpillars.

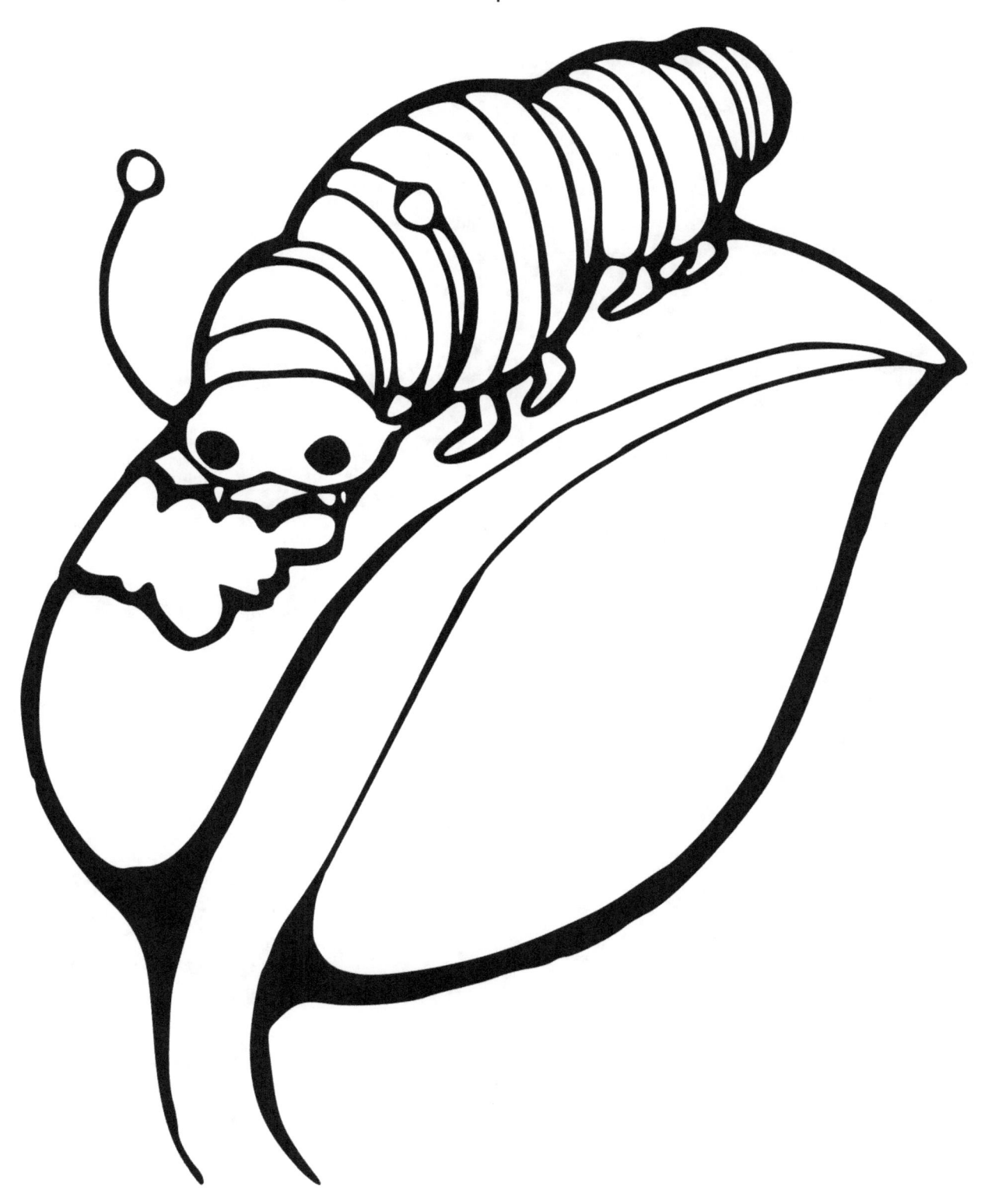

ADULT HORNETS

Adult hornets mostly drink plant juices

ADULT HORNETS

While larvae are fed protein, adult hornets mainly feed on nectar, sap, and sweet fruit juices, giving them the energy to fly and hunt.

HORNET ECONOMICS

The yellow-legged Asian hornet sometimes causes big money problems in France because it hurts bee colonies and people have to spend a lot to find and destroy its nests!

HORNET ECONOMICS

If the Asian giant hornet spreads across North America, it could mean loss of honey and crops.

HORNET ECONOMICS

In northern Spain, stopping the spread of the Asian giant hornet costs a lot of money!

LOCAL ECOSYSTEMS

Hornets are found in many countries around the world, including the United States, Canada, most of Europe, parts of Asia like Japan and China, and even some regions in Africa.

LOCAL ECOSYSTEMS

Hornets help crop rotation by eating pests that might attack different crops, making it easier for farmers to grow healthy plants year after year.

LOCAL ECOSYSTEMS

Hornets help recycle nutrients by breaking down dead insects and plants, which helps keep the soil rich and healthy for other living things.

LOCAL ECOSYSTEMS

The Asian hornet is a major problem for pollinators, especially honeybees.

Challenge: Do you know which country the Asian hornet (Vespa velutina), also known as the yellow-legged hornet, originally comes from?

LOCAL ECOSYSTEMS

European hornets, while large, are less harmful to bee populations and ecosystems.

LOCAL ECOSYSTEMS

In Africa, hornets live mainly in forests,
woodlands, and sometimes near farms or villages.

LOCAL ECOSYSTEMS

African honey bees are very good at protecting their home. If they get scared, many bees come out quickly to chase away danger.

LOCAL ECOSYSTEMS

The African honey bee has earned the nickname, "killer bees," but their sting is just like other bees. They just get angry faster!

LOCAL ECOSYSTEMS

African honey bees are native to Sub-Saharan Africa, but live in some places in the Americas now!

Challenge: Can you color five countries in sub-Saharan Africa?

LOCAL ECOSYSTEMS

Hornets build their nests in trees, bushes, or even inside walls, and when they leave, other small animals like spiders or beetles might move in to use the empty nest as a home.

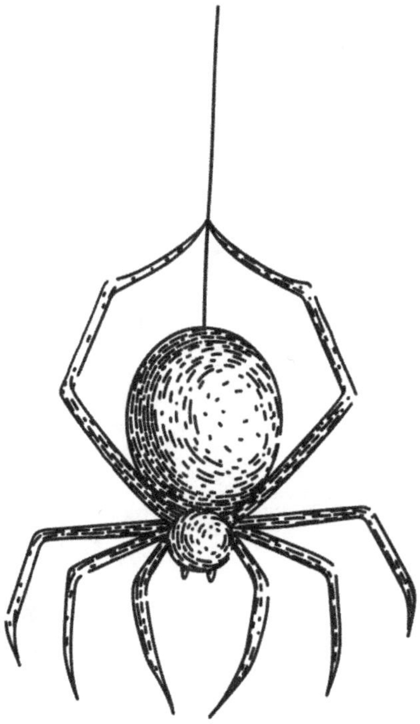

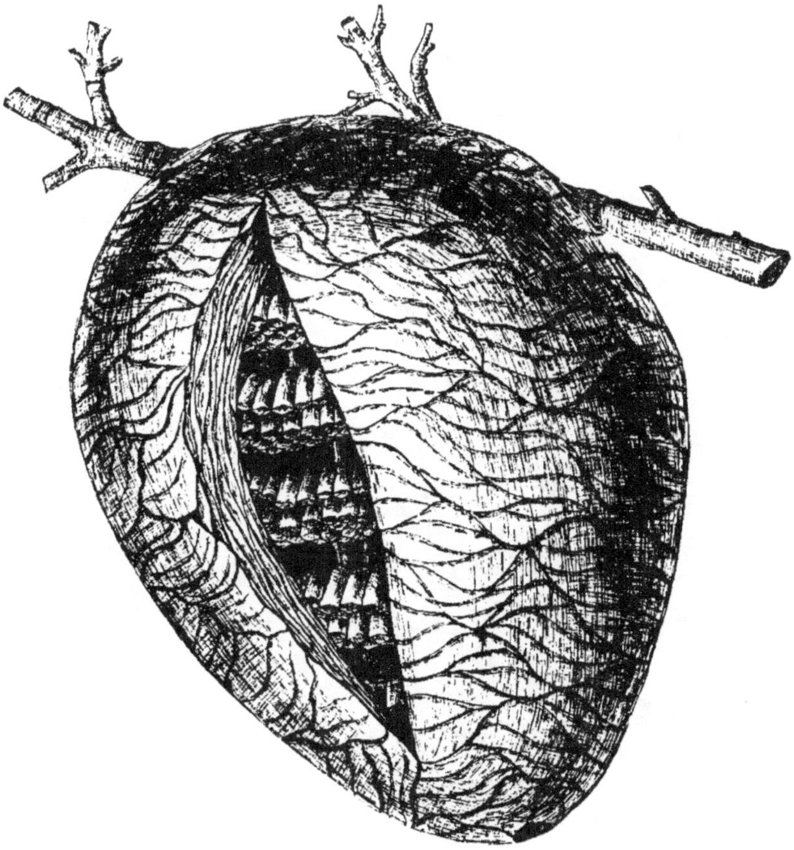

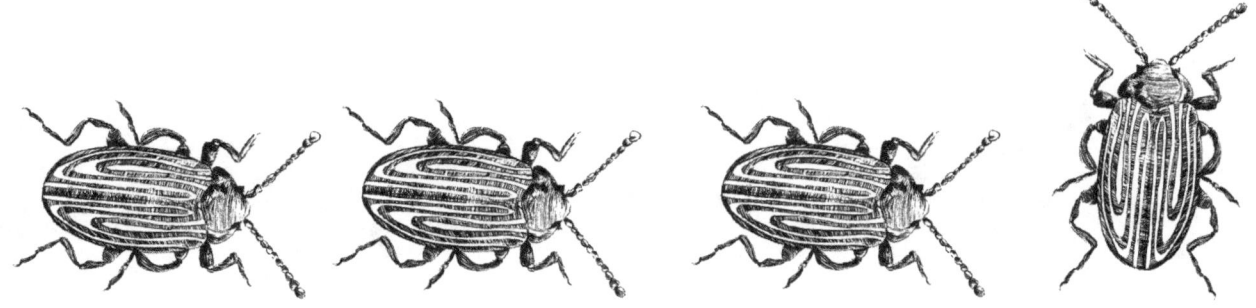

LOCAL ECOSYSTEMS

Asian giant hornet queens can grow as long as two inches — that's really big for an insect!

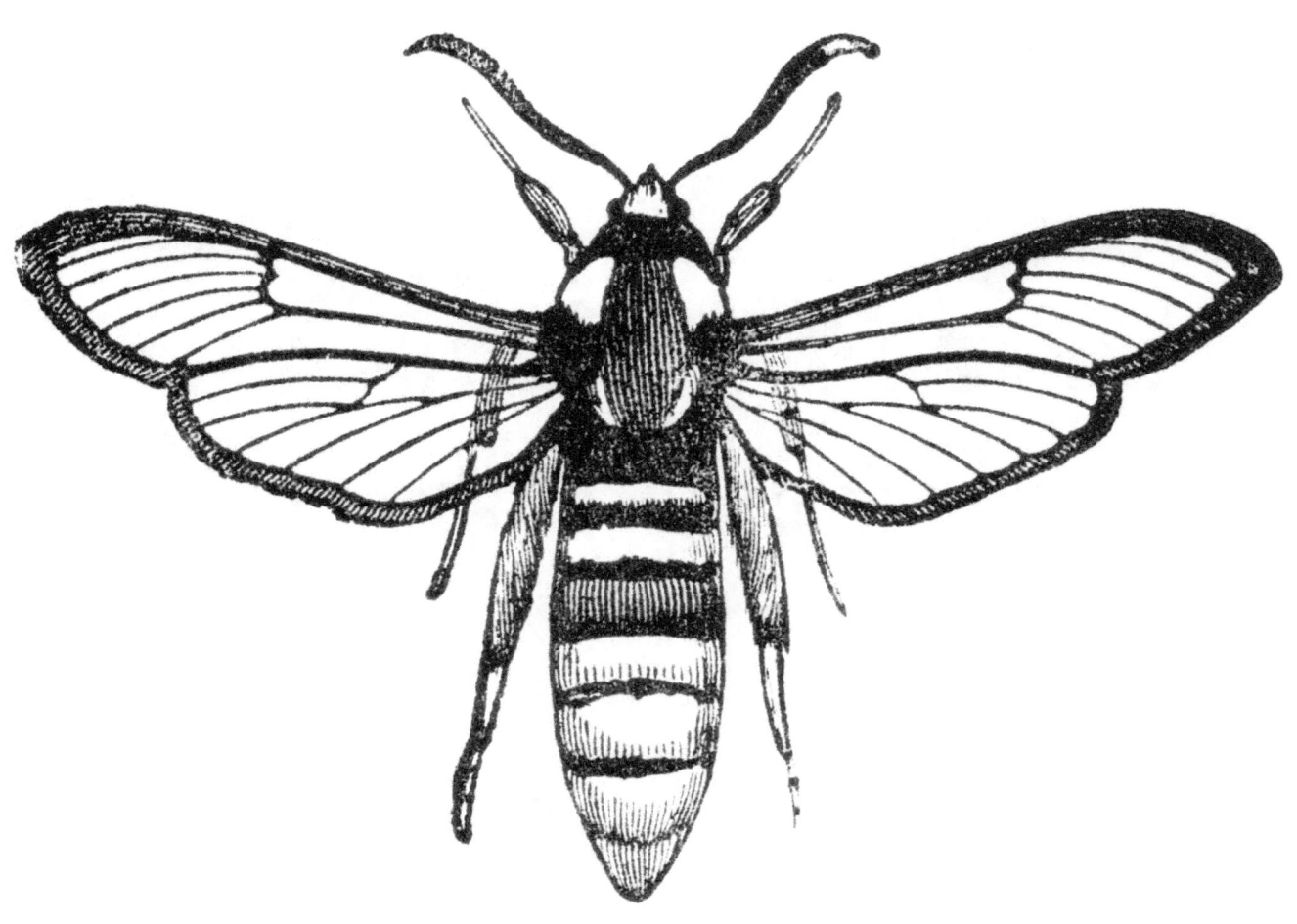

LOCAL ECOSYSTEMS

Asian giant hornets use sharp, saw-like jaws to attack honeybee hives and carry away parts to feed their babies.

LOCAL ECOSYSTEMS

The Asian giant hornet sting is very strong and painful - it can even go through special beekeeping suits!

LOCAL ECOSYSTEMS

Besides helping flowers grow, bees also support local ecosystems by providing food for animals like birds and frogs, and their leftover honey and wax help other insects and even plants thrive!

LOCAL ECOSYSTEMS

When bees visit flowers, they help plants produce seeds and fruits, which fall to the ground and grow into new plants.

LOCAL ECOSYSTEMS

Many wildflowers and native plants depend on bees. These plants create habitats and food for other animals, keeping the whole ecosystem strong.

BEES & HORNETS

Bees and Hornets each have five eyes! Three on top of their heads and two big ones on the sides.

BEES & HORNETS

Bees and Hornets are insect "cousins" - along with ants and sawflies!

BEES & HORNETS

Bees and Hornets both have wings and can fly.
Fast flyers with strong buzzing sounds!

BEES & HORNETS

Bees and Hornets live in colonies with a queen, workers, and a special job system. Teamwork helps their groups survive and grow.

BEES & HORNETS

Bees and Hornets can both sting (but only hornets can sting more than once). Bees lose their stinger after one sting, but hornets don't!

Stinger

BEES & HORNETS

Bees are amazing pollinators, and Hornets are great pest controllers.

NOTES